Contents

Any words appearing in the text in bold, **like this**, are explained in the glossary.

What are drugs?

A drug is any substance that affects your body and changes the way you feel. There are three groups of drugs – **medicines**, **legal drugs** and **illegal drugs**.

Medicines

Many medicines, such as cough medicines and painkillers, help to soothe the symptoms of a disease. Other medicines, such as antiseptic cream and antibiotics, tackle the disease itself. Some medicines can only be **prescribed** by doctors, but others can be bought from a chemist or from shops such as supermarkets.

This tablet contains the drug ibuprofen, a powerful painkiller. It is important to follow the directions on the box and not take more painkillers than are allowed, otherwise they can damage your liver.

Alcohol

Angela Royston

www.heinemann.co.uk
Visit our website to find out more information about **Heinemann Library** books.

To order:
 Phone 44 (0) 1865 888066
 Send a fax to 44 (0) 1865 314091
Visit the Heinemann Bookshop at www.heinemann.co.uk to browse our catalogue and order online.

First published in Great Britain by Heinemann Library,
Halley Court, Jordan Hill, Oxford OX2 8EJ
a division of Reed Educational and Professional Publishing Ltd.
Heinemann is a registered trademark of Reed Educational & Professional Publishing Ltd.

OXFORD MELBOURNE AUCKLAND
JOHANNESBURG BLANTYRE GABORONE
IBADAN PORTSMOUTH (NH) USA CHICAGO

Designed by AMR
Illustrations by Art Construction
Originated by Ambassador
Printed in Hong Kong/China

05 04 03 02 01
10 9 8 7 6 5 4 3 2 1

ISBN 0431 09907 3
This title is also available in a hardback library edition (ISBN 0 431 09902 2)

British Library Cataloguing in Publication Data
Royston, Angela
Alcohol. – (Learn to say no)
1.Alcohol – Physiological effect – Juvenile literature
2.Alcoholism – Juvenile literature
I.Title
362.2'92

Acknowledgements
The Publishers would like to thank the following for permission to reproduce photographs: Advertising archives, pp.20, 21; Gareth Boden, pp. 4, 7, 22, 24, 29; Image Bank, p.28 (Erik Simmons); Magnum, p.18 (Chris Steel-Perkins), p.23 (Abbas); Mary Evans Picture Library, p.9; Popperfoto, p.13; Rex Features, p.14; Science Photo Library, p. 17 (Andrew Syred); Telegraph Colour Library, p.19; Tony Stone, p. 5 (Leland Bobbe), p.6 (Paul Harris), p.25 (Zigy Kaluzny); Topham, p. 26.

Cover photograph reproduced with permission of Greg Evans and Telegraph Colour Library

Illegal or legal?

Legal drugs include medicines of course, but the term usually refers to drugs such as alcohol and tobacco. These drugs affect the way a person feels but are not illegal for adults. Tea, coffee and cola are legal drugs too. Illegal drugs include **cannabis**, heroin, Ecstasy and LSD and are forbidden by law.

Alcohol

This book will tell you about alcohol – how it affects both the body and the way in which a person behaves. It looks at why people drink and what can happen when they drink too much. It gives you plenty of reasons why you should avoid drinking alcohol and explains the dangers that can result from drinking alcohol or getting **drunk**.

Did you know?

Alcohol is a **depressant.** Apart from medicines, drugs are depressants, **stimulants** or **hallucinogens.** A depressant makes a person relax. A stimulant, such as caffeine in coffee or nicotine in tobacco, makes your heart and other organs work faster. A hallucinogen, such as cannabis, alters the way a person sees or hears things.

Coffee, tea and cola all contain caffeine, a legal drug which makes the heart beat faster and so makes you feel more awake.

What is alcohol?

There are several kinds of alcohol, including methanol and ethanol. Methanol and ethanol, made from crude oil, are used in industry. They are used as a **solvent** in varnishes, stains and lacquers and added to petrol to make the fuel work better. Poisonous chemicals are added to industrial alcohol so that no one can drink it. The alcohol in **alcoholic drinks** is ethanol and is made from plants.

Alcohol from plants

Almost any plant can be made into alcohol, but most alcoholic drinks that are produced commercially are made from grapes or cereal crops. Grapes are made into wine, sherry and brandy. Cereal grains, such as barley and rice, are made into beer, lager, whisky and gin. Vodka is made from a mixture of grain, sugar beet and potatoes. Rum is made from sugar cane.

Beer and lager are made in a **brewery**. **Yeast** is added to barley, hops and water and the mixture is left to ferment.

Fermentation

Plants have to **ferment** before they produce alcohol. They are mixed with water and a little yeast is added. The yeast slowly changes the sugar in the plant mixture into alcohol. When the fermentation is complete, the liquid is filtered and left to age for several months or even, in the case of wine, for several years. As it ages, the drink develops its particular taste.

Spirits

Some alcoholic drinks are stronger than others. Beer and wine contain a lot of water, so are not as strong as other drinks. **Spirits**, such as whisky, vodka and rum, are distilled to make them stronger. Distilling separates and removes more of the water, making the alcohol more concentrated. Because they are stronger, spirits make people **drunk** more quickly. Many people add water or **non-alcoholic drinks** such as cola or lemonade to spirits to make them weaker.

Did you know?

The strength of an alcoholic drink is measured as a percentage. A drink that is 50-per-cent (one half) alcohol is said to be 100 **proof**. These are the approximate strengths of some drinks:

Beer	5%
Wine	10%
Sherry	20%
Spirits	45%

These bottles all contain alcohol. Spirits, such as whisky, gin, brandy and vodka, are stronger than beer or wine, and are more expensive.

A long tradition

Alcohol was probably first made by accident, when grain or grapes were stored and began to **ferment**. No one knows who first made alcohol, but we do know that beer was made thousands of years ago by the earliest civilizations. Designs on pottery made in Mesopotamia over 6000 years ago show alcohol being fermented.

Drinking in the past

From the earliest times people drank alcohol on special occasions. The Ancient Egyptians made wine and in the days of the Roman Empire wine was drunk during special feasts that lasted for days. Brandy was being distilled over 2000 years ago and is probably the oldest form of **spirits**.

Drinking problems

Drinking a glass or two of alcohol from time to time is not harmful to adults, but some people drink too much. Then they put themselves and other people in danger. Some people become **addicted** to alcohol – they cannot manage without large amounts of it almost every day and find it hard to stop drinking.

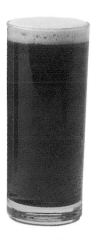

Monks in the Middle Ages used to make their own wine or mead in the monastery. Today, beer companies use names, like Abbot Ale, to give their product a traditional feel.

A history of trouble

During the 19th century, millions of people in Europe and America were living in city slums and working for low wages in factories where conditions were harsh. In despair, many people started to drink heavily and drunkenness became a major problem. Some religious groups formed **temperance societies** to persuade everyone to stop drinking.

During the 1700s many people drank too much. English painter, William Hogarth (1697–1764) drew this picture of life in a London street.

Alcohol today

Today drunkenness still causes huge problems. Over half of all violent assaults within the home are committed by someone who has been drinking alcohol. **Drunk** youths get involved in street fights and pub brawls and many motor accidents are caused by drunk drivers.

Did you know?

Alcohol deadens the body's reactions. In the 1800s, before better **anaesthetics** were discovered, alcohol was used as an anaesthetic. Before having an operation, some patients were given so much alcohol that they passed out, or lost consciousness. However, the pain of the operation quickly brought them round.

Alcohol and the law

Alcohol can be a dangerous drug. It is particularly dangerous for children to drink alcohol because their bodies absorb it more easily. It takes much less alcohol to make a child **drunk** than an adult drunk. In Britain and Australia and most other countries there are laws which forbid shops, bars and restaurants from selling alcohol to young people. It is also illegal to sell alcohol to someone who is buying it for a young person.

The law

In Britain, 16-year-olds can drink wine or beer, but not **spirits**, with a meal in a restaurant. They cannot drink in a bar or buy alcohol until they are 18. The law varies from state to state in the United States and in Australia, and from country to country in Europe. But the message is usually the same – children should not drink.

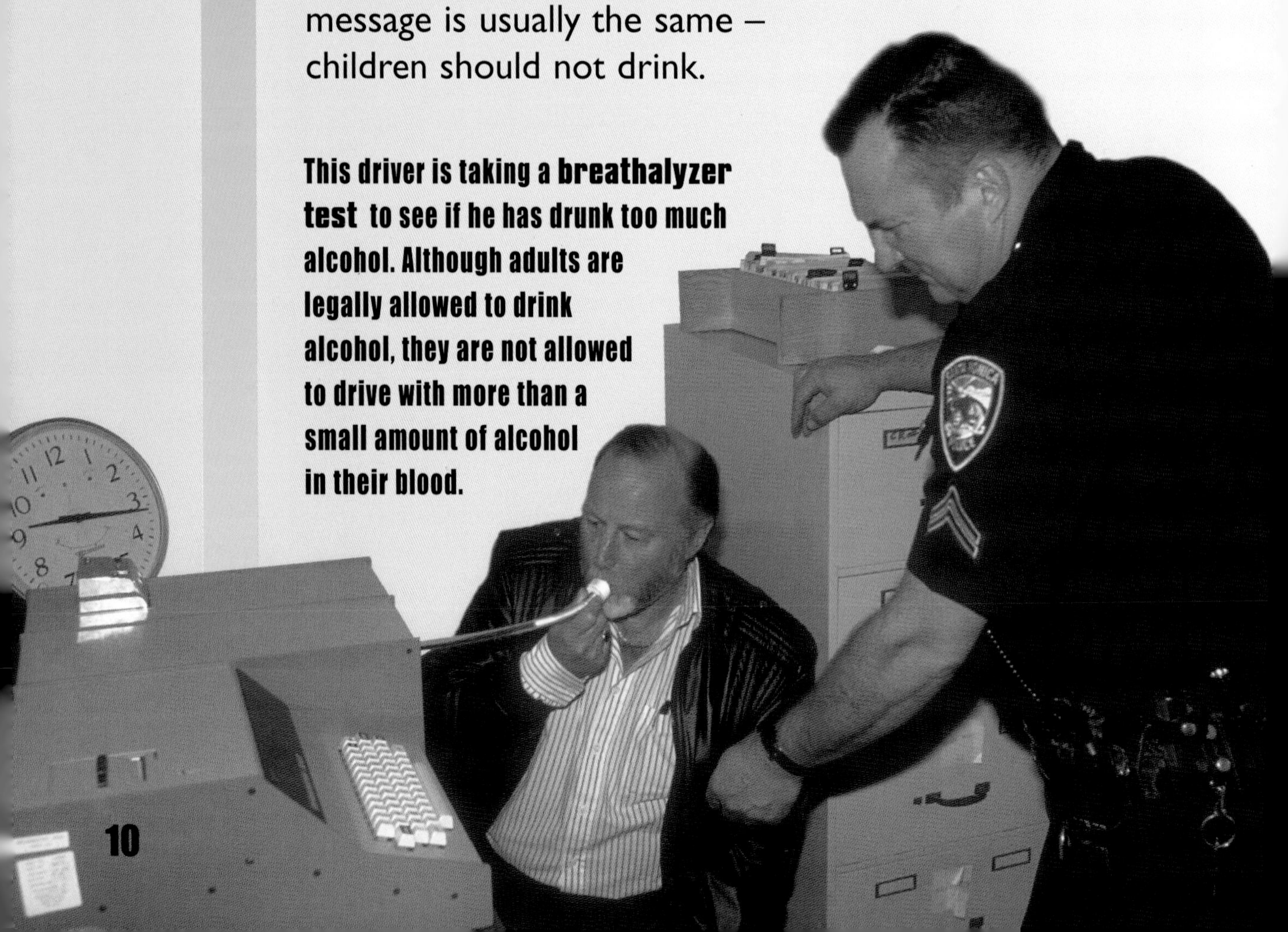

This driver is taking a breathalyzer test to see if he has drunk too much alcohol. Although adults are legally allowed to drink alcohol, they are not allowed to drive with more than a small amount of alcohol in their blood.

Penalties

There are strict penalties for people who break the law. Alcohol cannot be sold just anywhere. A shop, bar or restaurant has to apply for a **licence** to sell it. If someone is caught selling alcohol to a person who is **under age**, the owners of the premises could lose their licence and the person their job. Bar staff may ask young people to prove that they are over 18 before selling alcohol to them.

This family in Baghdad, Iraq are not allowed to drink alcohol. Several Muslim countries ban alcohol completely.

Responsible age

There is no particular age at which the law decides you become an adult. You are legally allowed to marry at 16, but are not allowed to vote or buy alcohol until you are 18. Do you think it is right that 16-year-olds can marry but not buy wine for their wedding, or do you think that 16 is too young to marry anyway?

What happens when someone drinks?

Effects on the body

Alcohol is a poison that the body tries to get rid of. The liver breaks alcohol down into harmless sugar, but the liver can only deal with less than 15 grams of alcohol an hour. Any extra alcohol stays in the blood and is taken around the body.

If the person goes on drinking, the alcohol builds up in the blood. A little of it is breathed out, which is why you can always smell alcohol on someone's breath if they have been drinking. Some of the alcohol is removed by the kidneys and passes into the urine. If a person drinks a lot of alcohol, it may make them sick. This is the body's quickest way of getting rid of any poison. It can take several hours for the effects of alcohol to wear off.

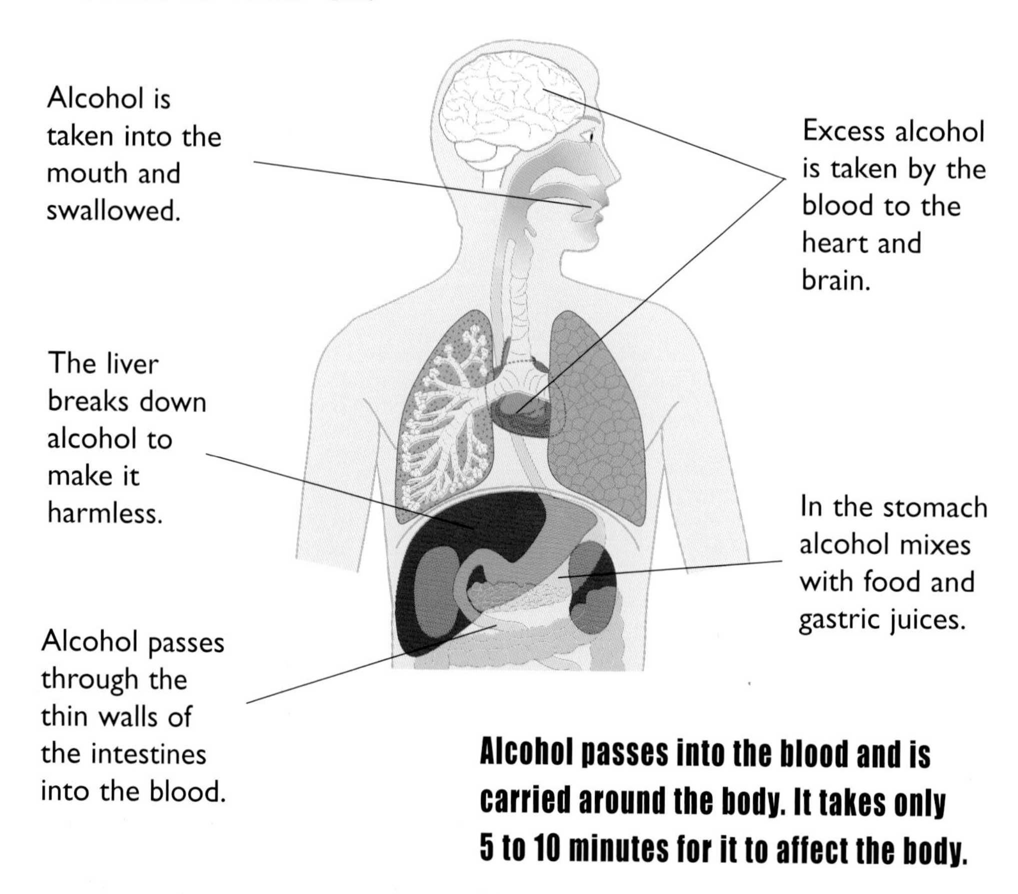

Alcohol is taken into the mouth and swallowed.
Excess alcohol is taken by the blood to the heart and brain.
The liver breaks down alcohol to make it harmless.
In the stomach alcohol mixes with food and gastric juices.
Alcohol passes through the thin walls of the intestines into the blood.

Alcohol passes into the blood and is carried around the body. It takes only 5 to 10 minutes for it to affect the body.

Hangovers

Someone who has drunk too much may suffer from a **hangover** the next day. They may have a bad headache, dry mouth, sore eyes and feel sick. One of the main causes of a hangover is lack of water. Drinking alcohol actually makes the body lose water and become **dehydrated**. It can take several hours to recover from a hangover.

This man has probably passed out from drinking too much alcohol. He is unaware of what is happening to him. Other people also seem to be unaware that he may be seriously ill or have injured himself when he fell.

Did you know?

Someone who drinks too much alcohol may pass out and may die from alcohol poisoning. Every year in Britain 1000 people under the age of 15 are taken to hospital with alcohol poisoning.

Effects on behaviour

Some adults say that they like to drink alcohol because it makes them relax and helps them to enjoy themselves. Jokes may seem funnier than they really are. People say that they feel less inhibited, which means that they are less cautious about how they behave. Alcohol can help a shy person to feel more confident. But it makes some people more aggressive. Being **drunk** affects a person's judgement and ability to make decisions.

Loss of control

Different people are affected by different amounts of alcohol. But scientific tests show that after just one or two drinks everyone is less able to do intricate tasks. They find it more difficult to balance small objects on top of each other, for example, or do mental arithmetic sums. After several drinks a person begins to slur their words and lose their balance. Women usually get drunk quicker than men, but the people who are most affected are **under age** drinkers.

When people are drunk, they often do not realize that they are behaving differently from usual. And if they do realize it, they usually don't care!

Bad judgement

The most dangerous thing about being drunk is that judgement is affected. A drunk person is more likely to have an accident. Each year, a third of pedestrians killed in accidents have been drinking. Children who have been drinking are even more likely to act without thinking. When people are drunk, they are more likely to get carried away and may find themselves doing things, such as having sex, even if they had not planned to.

Did you know?

Alcohol kills brain cells. Brain cells do not renew themselves in the way that other body cells do. Millions of nerve cells make up the brain, and some of them die naturally during your life. Drinking alcohol speeds up the process.

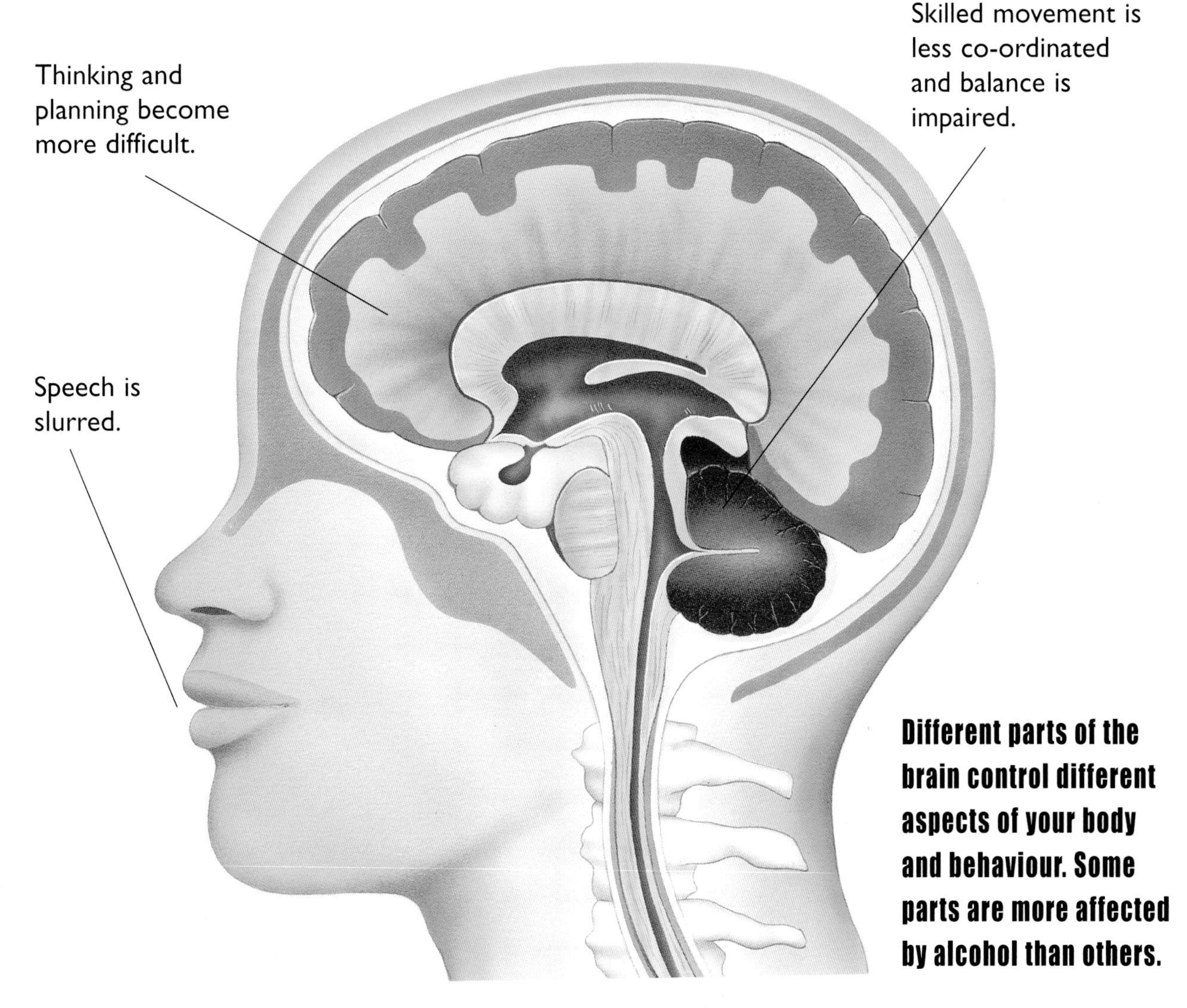

Thinking and planning become more difficult.
Speech is slurred.
Skilled movement is less co-ordinated and balance is impaired.

Different parts of the brain control different aspects of your body and behaviour. Some parts are more affected by alcohol than others.

Frequent heavy drinking

Alcoholism

The more a person regularly drinks, the more they need to drink before they become **drunk**. Some people get so used to drinking a lot of alcohol that they become an **alcoholic**.

Alcoholism affects people in different ways. Some alcoholics drink all day, from the time they get up until they fall into a drunken slumber. Others rely on drink to keep them going, but may not drink every day. Many alcoholics can stop drinking for a certain amount of time, but go back to it after a few weeks.

Effects on the body

People who drink heavily often become overweight and underfed. In the body, alcohol is changed to sugar which gives the body the energy it would normally get from food such as bread, pasta, potatoes and rice. The unused energy is stored in the body as fat. One or two drinks can make people feel hungrier and so they eat more. More alcohol, however, stops the feelings of hunger so many heavy drinkers do not eat enough.

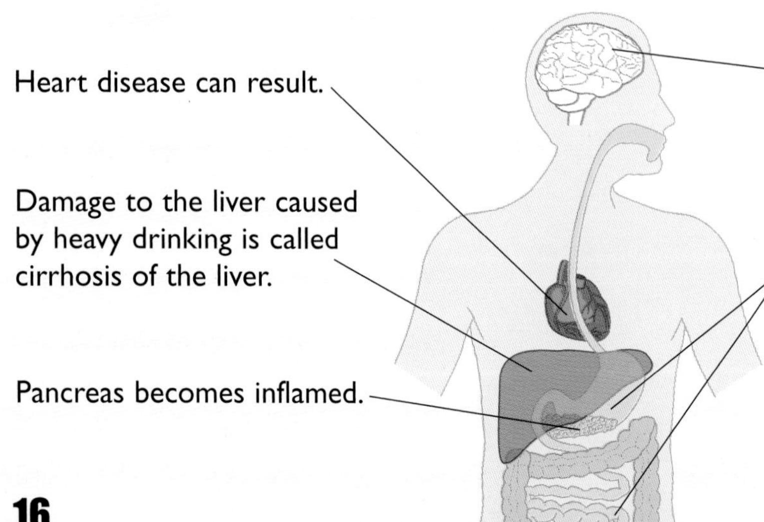

Heart disease can result.
Damage to the liver caused by heavy drinking is called cirrhosis of the liver.
Pancreas becomes inflamed.
Brain damage. One sign of brain damage is memory blackouts (periods of time that the person cannot remember).
Alcohol irritates the digestive system and can cause ulcers and cancer.

Very heavy drinking over several years can damage the body in several ways.

Serious diseases

The liver and the digestive system are most affected by alcohol. The lining of the throat, food pipe and stomach become inflamed and, over a long time, this can cause serious diseases. Over-drinking has been linked to some cancers, heart disease and strokes. In Britain, up to 28,000 people die each year from diseases related to alcohol – 50 times more than the number of deaths caused by **illegal drugs**.

Did you know?

In the United States 10 million people are thought to be alcoholics. People of all ages can be alcoholic, but most alcoholics (about three-quarters) are men. Alcoholism is also common in France, Ireland, Poland, Scandinavia and Russia.

The liver filters out poisons, such as alcohol, from the blood and changes them into harmless substances. This liver, however, has been damaged by having to process too much alcohol over a long period of time.

Long-term effects on behaviour

People may start to drink heavily to get away from other problems, but drinking soon makes these worse. Drinking is expensive and will make money problems worse. Heavy drinkers are often late for work. They may lose their job, particularly if they are found to be **drunk** or drinking at work.

Some heavy drinkers not only lose their jobs, but also their homes and families. They end up living on the street, unable to give up drinking.

Family and children

The families of **alcoholics** suffer as well. If family money is being spent on alcohol, there will be less to spend on clothes, holidays and even food for the children. A parent cannot look after children properly when they are drunk. Often, the children end up looking after the parent and trying to hide their drinking from other people.

Violence

Many people become violent when they are drunk. They may be controlled and reasonable when they are sober, but when they have had a few drinks they become aggressive. Many men who beat and ill-treat their partners and children only do so when they are drunk. Someone who becomes aggressive when they are drunk may start fighting in the street or bar. Drunken brawls can lead to fights in which people are seriously injured.

Did you know?

About half of all violent assaults, both at home and in public, are done by men who are drunk. Nearly a third of murders are committed by people who have been drinking and two-thirds of suicides are related to heavy drinking.

Nearly nine out of every ten people arrested for damaging property have been drinking alcohol.

Helping an alcoholic

Some people say that when the family of an alcoholic is sympathetic and put up with their drinking, they are **colluding** with the alcoholic. They say that the alcoholic will have to face the consequences of their drinking before he or she finds the will to recover. Do you think this makes sense or do you think it is too harsh?

Why do people do it?

Thinking for yourself

Young people may decide to drink for several reasons. They may be curious or their friends may be doing it and they don't want to be left out. However, it is important to think for yourself and not just follow what other people do. If one person in a group says no, all the others who are doubtful may say no too.

A 'cool' image?

Some children think it is 'cool' to drink. They don't realize that getting **drunk** is not only dangerous but can make them look stupid. Young people often drink as a way of rebelling and to show that they can make their own decisions. But adults are more likely to let young people be independent if they know they will be sensible.

This advertisement wants you to think that drinking alcohol is cool and sophisticated. The companies who produce alcohol spend a lot of money finding out why people drink, then they produce advertisements that are most likely to appeal to them.

Why adults drink

Some adults drink too much alcohol. They may go to a bar to celebrate a birthday, for example, but end up drinking too much. Other people may drink to get away from problems in their lives. Perhaps they are frustrated in their work or they cannot find a job. Some people drink because they feel bad about themselves. Drinking, they think, helps them to forget about their problems. But drinking solves nothing and may lead to **alcoholism**.

Did you know?

Some people drink to cheer themselves up, but alcohol tends to make people depressed. If someone starts drinking because they are miserable they will probably end up feeling much worse.

What kind of situation does this advertisement from the 1970s use to sell its lager?

Warning labels

Cigarette packets and advertisements for tobacco have to carry a government health warning which tells people that smoking causes cancer and damages their health. Do you think that alcoholic drinks should also carry a warning? If so, what do you think it should say?

Saying no

Whether you eventually decide to drink alcohol or not, you will come across people who try to persuade you to drink when you do not want to. Even adults are urged to have a drink, or to have another one, when they have already said no. What will you do when this happens to you?

How to say no

Many people are worried that if they say no, their friends will think they are stupid and childish. But it is more mature to be able to say you don't want to drink than to accept a drink that you don't want. This book gives you plenty of reasons for saying 'no', such as 'I don't like feeling out of control'. You can think of other possible things that you might say, such as 'I still have to do my homework' or 'I'm not allowed to drink'. But you don't have to give an excuse.

You don't have to drink alcohol to have fun. In fact, drinking alcohol when you are young will spoil your fun .

Don't hesitate

What you say matters less than how you say it. If you are sure you don't want the drink, then say so clearly. Friends are more likely to try to persuade you if they sense that you are unsure.

Did you know?

In the 1920s it was illegal to make or sell alcohol anywhere in the United States, but illegal alcohol was still available. While this law (called Prohibition) lasted, dangerous gangsters such as Al Capone sold alcohol illegally to make money.

Muslims are forbidden by their religion to drink alcohol. These wedding guests do not need alcohol to help them enjoy themselves. Many Christian sects also forbid alcohol.

Licence to drink

How well do government controls on the sale of alcohol work? Do you think that the sale of alcohol should be more controlled than it is? Or do you think that banning the sale of alcohol leads to unnecessary crime?

Avoiding difficult situations

Try to avoid situations where you might be asked to drink when you do not want to. For example, if you suspect that a classmate is planning to raid her parent's drinks cupboard and asks you to the 'party', it is better to find an excuse not to go, than to go and try to not drink.

Many young people like to spend time with their friends after school. If your friends use this time to drink alcohol, it is better to make an excuse and go home. You could always take up another interest, such as playing a sport or joining an after-school club.

Spending time with your friends after school is important, but if they are planning to do something you don't want to, it may be wiser to go straight home.

Making new friends

If your friends are experimenting with alcohol, cigarettes or other things you do not want to try with them, you should think about whether they are the best friends for you. Friends should be people who respect your views and accept you for who you are. Do not let your friends try to make you more like themselves.

New pursuits

Friendships are important and it is not always easy to change friends. Nevertheless, it may be possible to widen your circle of friends and spend time doing new things that you will enjoy even more.

Did you know?

For the price of a 250 ml can of lager, you could buy a 1.5 litre bottle of fizzy drink. For the price of a bottle of whisky, you could buy a CD or two huge pizzas.

There are plenty of clubs and activities that you can do after school. You may make new friends this way too.

Giving up

If someone is an **alcoholic**, giving up drinking is not easy. The first step is for the person to recognize that they have a drinking problem. The second step is to seek help from an organization such as Alcoholics Anonymous. The alcoholic's family often need to be helped too.

Alcoholics meet together and talk about their drinking problems. They help each other to stop drinking.

Road to recovery

Alcoholism is a disease from which people can recover but cannot be cured. People do stop drinking and can get their lives back together again. But they cannot risk drinking any alcohol. One drink could be enough to start them drinking again.

Sensible drinking for adults

Most people who drink a lot are not alcoholics, but heavy drinking is unhealthy. Doctors say that one or two drinks are all right for adults, but they stress that drinking should be moderate and sensible. They measure alcohol in terms of **units**. No more than a few units should be drunk at one time and no more than a certain number should be drunk in a week.

Did you know?

Doctors measure alcohol according to units. A glass of wine, a can of beer or lager and a measure of **spirits** are all one unit. Men should not drink more than 28 units a week. Women should not drink more than 21 units.

Is it legal?

The UK has laws about how old a person can be before they are allowed to buy or drink alcohol.

- Young people under 14 years old cannot go into the bar of a pub unless it has a 'children's certificate'. They can only go into parts of **licensed** premises where alcohol is either drunk but not sold, for example a pub garden, or where alcohol is sold but not drunk, for example a sales point for alcohol to be drunk away from the pub.
- Young people aged between 14 and 15 years old can go anywhere in a pub but not drink alcohol.
- Young people aged between 16 and 17 years old can drink beer or cider with a meal but not in a bar.
- It is against the law for anyone under 18 years old to buy alcohol (except for 16 or 17 year olds having a meal in a pub).
- It is illegal for anyone to buy alcohol in a pub for someone under 18 years old.

Do you think these laws are fair? Why do you think there are different rules for people of different ages?

Dealing with problems

Difficult times

Growing up is an exciting and enjoyable time. It is a time when you can try out new things and begin to make decisions for yourself. It can also be a difficult time. There are important exams that you have to work hard for. Friendships with your own sex and the opposite sex won't always run smoothly. Most young people worry about the way they look, whether people will like them and what they are going to do with their lives.

Finding the right solution

Don't think that drinking will make your life better. If you are happy and want to have a good time with your friends, you can do so without getting **drunk**. If you are worried or miserable, drinking will not help you. It is better to face up to your problems and try to sort them out.

Talking over a problem can help you feel better or help you to find a solution. Try talking to a friend you can trust or someone in your family.

Talking it over

If you have problems that you feel you cannot deal with, don't despair. There are people and organizations that can help you. If you can, ask for help from your family, your friends or your teachers. If at first the people you talk to don't listen to you, go on looking and asking for help until you do get it.

Did you know?

Childline is an organization in Britain that provides help and support for children and young people in trouble or in danger. The number is 0800 1111.

If you cannot talk to someone in your family, you can talk to another adult you trust, such as a teacher you like or a youth club leader.

Useful contacts

Alcohol

Al-Anon – information on self-help groups for adults, families and friends of alcoholics:
telephone: 020 7403 0888.

Alateen – information on self-help groups for young people aged 12–20:
telephone: 020 7403 0888.

Alcoholics Anonymous – for information about local self-help groups and local helpline, write to:
AA, PO Box 1, Stonebow House, Stonebow,
York YO1 2NJ
or telephone: 01904 644026 Monday-Thursday 9.00 am to 5.00 pm, Friday 9.00 am to 4.30 pm.

Australian Drug foundation – for information write to:
409 King Street, Melbourne 3000, Australia
or telephone: 03 9278 8100
or website: www.adf.org.au

Childline – provides free support for children or young people in trouble or danger:
freephone helpline: 0800 1111.

Drink helpline – for information and advice on problems related to alcohol:
freephone: 0800 917 8282
open Monday-Friday 09.00 am to 11.00 pm; Saturday-Sunday 6.00 pm to 11 pm.

Drug info Line –
telephone: Victoria 131570 or NSW 02 93612111

Direct Line – for counselling:
telephone: Australia 1800 136 385

Glossary

addicted unable to give up a habit. Alcohol becomes addictive when a person cannot manage without it.

alcoholic addicted to alcohol

alcoholic drinks drinks which contain alcohol

alcoholism disease in which a person is addicted to alcohol

anaesthetic substance which stops the feeling of pain

breathalyzer test a test which measures the amount of alcohol in a person's breath and therefore shows how much alcohol is in the blood

brewery factory where beer and lager is brewed

cannabis an illegal drug made from hemp. Cannabis is also called marijuana and hashish.

cell the building block of all living things, including the human body

dehydrated lacking in water, thirsty

depressant a substance that slows down the body's reactions and relaxes the muscles

drunk strongly affected by alcohol

ferment change into alcohol with the help of yeast

hallucinogen a drug that heightens experiences or makes things which are imagined seem real

hangover unpleasant physical affects experienced after being drunk

illegal drug a drug, such as heroin, LSD or cannabis, which is forbidden by law

legal drug a substance which affects the body but is allowed by law. Medicines, coffee and tea are legal drugs.

licence a legal permit

medicine substance which is used to treat or cure illnesses

non-alcoholic drinks drinks which contain no alcohol

prescribed given a medicine under the advice or order of a doctor

proof a measure of the amount of alcohol in a drink

solvent a liquid in which another substance is dissolved

spirits alcoholic drinks which have been made more alcoholic by being distilled, that is, having some of the water in the drink removed

stimulant a substance that speeds up the body

temperance societies groups of people who want to persuade everyone not to drink alcohol

under age younger than the legally allowed age

unit A way of comparing the amount of alcohol in different drinks. A glass of wine, a can of beer or a measure of spirits are all one unit.

yeast a substance that helps sugar in plants to be changed into alcohol

Index